C0-ARU-639

HUMAN NATURE

HUMAN NATURE

Sustainable Farming in the Pacific Northwest

ANNA MIA DAVIDSON

For Amina,
with love and appreciation!
from Mia

With Essays by

SEBASTIÃO SALGADO

MATT DILLON

DR. MARCIA OSTROM

minor matters.

CONTENTS

Page 2: *Oliva with Chicken* (detail), Frog Song Farm, Fir Island

Opposite: *John on Tractor* (detail), Nature's Last Stand, Fall City

ALLIED

Introduction

SEBASTIÃO SALGADO

Anna Mia Davidson's choice—to return to nature—is close to my heart, as is her decision to share her experience of sustainable farming across the Pacific Northwest in this splendid collection of photographs. On a personal note, I get a special thrill from knowing that the little girl I knew at her parents' home in New York many years ago is now the fine and committed artist of these pages.

At first view, many of these images might seem to be portraits: of her hard-working husband John, her children, and neighboring farmers, as well as of chickens, pigs, potatoes, and lettuce. But they tell a far broader story. They remind us that by respecting animals, plants, and the very soil of the earth, we are contributing to our own survival.

Anna Mia Davidson is lucky, but luck came to her only because she and John made the daring choice of turning away from one way of life and embracing another. In one sense, they have stepped back to almost a pre-industrial rural existence, like homesteaders of a different age. But in another sense, they are reaching out for a different future. Sustainable farming should not be the privileged option of only developed countries. It is an absolute necessity for the hundreds of millions of peasants living in rural poverty across the world.

In my travels, I have witnessed ecological devastation on a massive scale, from deforestation and industrial farming to poisoned rivers and skies darkened by heavy industry. But I am also lucky to have seen nature in its pristine form, in majestic mountains and volcanoes, in silent deserts and wild rivers, in untamed animals and friendly whales. And in the Amazon rainforest, the heights of New Guinea, and the savannas of Africa, I have watched how ancient peoples have developed their own versions of sustainable farming by balancing their needs with those of the nature that sustains them.

That's why Anna Mia Davidson's beautiful photographs touch me. They exude the pride of those who feel at one with their environment. They speak to me of the simple—but undoubtedly tough—lives that the sustainable farmers of the Pacific Northwest have adopted. Their objective, their success, is to produce organic food untainted by the industrial products that we unknowingly consume daily at breakfast, lunch, and dinner. But their method is one that conserves the gifts that nature has loaned them.

I believe it is possible to turn back the clock, because I have seen it happen in a property that my wife, Lélia Wanick Salgado, and I inherited from my parents in the Brazilian state of Minas Gerais. Decades ago, its trees were cut down to make way for cattle; over time, erosion turned the land into something of a desert. Then, in what at the time seemed like reaching for the impossible, we decided to reforest the property with the same species of the Atlantic Forest that once flourished there. Today, the slopes and valley of the property are again thick with trees; water has returned to streams;

and animal life—birds and insects, snakes, and jaguars—has again found a home there.

Perhaps, as much as my travels, this experience reinforced my faith in nature. We are told that humans are the only rational beings, yet reason is no less present in nature. Animals, plants, and trees all have their reasons to grow and multiply. To stand waist-deep in a fast-flowing river is to know its purpose—literally, its *raison d'être*—is to create and preserve life. Even mountains are alive—some rising up as the earth releases its energy through volcanic eruptions, other shrinking through erosion and landslides.

We cannot all experience this firsthand. Nor can we all choose—or do we all even want—to become sustainable farmers. But we owe it to ourselves to come closer to nature. I recognize that perhaps 99 percent of city dwellers see no alternative to their way of life, but I also believe that those same people can learn to understand and respect nature. It simply means opening one's eyes and mind to the mysterious world that surrounds us. And, in doing so, being richly rewarded in both body and soul.

Yes, Davidson and her family *are* lucky, because it seems evident from this book that they and their like-minded neighbors are happy with their choice. They may not be able to feed entire cities with the results of their labor. Ultimately, they offer one pioneering answer—not *the* global solution—to industrial farming. But when we next see organic fruit and vegetables on sale in our shops, *Human Nature: Sustainable Farming in the Pacific Northwest* will immediately come to mind as a vivid example of living in harmony with nature.

Cast Your Vote

MATT DILLON

Someone decided that factories are the more dynamic way to put food into our bellies—quicker, faster, cheaper—all the while tricking us into thinking that the *farm* is still at the core of the American ideal.

A factory outside with robot seeds that miraculously fight off pests and disease and grow to their full potential no matter what the season or circumstances?

That

is

not

a

farm.

It is a factory. Fueled by profit. And what it produces is not food.

Supporting local agriculture is not just about deliciousness. It's not just about restaurants or businesses using a tag phrase like "Farm to Table." It's also not just about the perfect glistening carrot on the farmers market stand.

It *is* about participation.

Deliciousness is one of the rewards of this participation, but surely not the only one. Environmental stewardship, health of our bodies, health of our communities, diversity, and economic growth are all things we gain by participating in our local agricultural community.

I remember when it hit me that—as a chef with a plate of food—I had the ability to impact more than just a guest in a restaurant. By looking beyond deliciousness to how I participated in the community around me *and* how it participated with me, I opened new doors. I understood that my choice to purchase with integrity, purpose, and a focus on farmers who wanted to be as close to me as I wanted to be with them gave me the ability to strengthen a bond that seems to be very weak in our country at the moment. The farmers I worked with became a part of my extended family. Almost twenty years later, many of them are still part of my routine and in my life. We share a bond that has been part of the human experience for ages.

I know it is hard—seemingly impossible—for everyone to be able to go to the farmers market or join a CSA [*see page 64*]. I know it is expensive. I know that there is a *huge* demographic of people for whom it is very, very difficult to participate in this approach. This is a gigantic socioeconomic problem that I have a hard time wrapping my head around.

It pains me to know that a small farmer, who makes so little, still has to charge a hefty (yet fair) price for a bunch of carrots. But, knowing that, I get to vote with my dollar every time I buy a carrot from a family farmer who is sharing in the world with me.

I get to vote NO, I don't want monoculture, period. No, I don't want my children to have five eyeballs or three ears. No, I don't want the bees or the butterflies or the salmon to go away. No, I don't want my small farmer friend to be broke. No, I don't want food factories posing as farms around anymore, giving a false economic sense of what it costs to grow a carrot. Particularly when after test tubes and chemicals and genetic modifications, it's not even a f*#king carrot. No.

Vote. With your dollar. On a carrot from a small, local farmer. Participate. For your neighbor. For your community. For yourself.

NATURE'S LAST STAND, FALL CITY

Nestled between the Snoqualmie River and Patterson Creek, this small-scale sustainable family farm owned by John Huschle and Anna Mia Davidson is rooted in the beliefs of conservation. Cultivating five of the owners' twenty-three acres in the Snoqualmie Valley, the farm produces vegetables, pasture-raised pigs, and egg-laying chickens.

John with Tractor at Dusk, Nature's Last Stand, Fall City

 Eden with Goldenrods, Nature's Last Stand, Fall City

Potato Harvest, Nature's Last Stand, Fall City

 Pierre Harvesting, Nature's Last Stand, Fall City

Tom Willis of Clean Greens Farm, Nature's Last Stand, Fall City

Pierre with Potatoes, Nature's Last Stand, Fall City

SUMMER RUN FARM, CARNATION

Summer Run, founded in 2001, is a Certified Naturally Grown farm (a grassroots movement made up of farmers who respect, implement, and even helped write the original national organic standards). The twenty-acre farm is owned by Cathryn Baerwald, who cultivates five acres of vegetables.

Cathryn on Tractor with Dog, Summer Run Farm, Carnation

Berry Pickers, Sidhu Farms, Puyallup
Left to right: Mr. Ghuman, Sharigit,
Mrs. Ghuman, Amarjit Kaur

SIDHU FARMS, PUYALLUP

Comprised of approximately ninety acres of plots throughout the Puyallup Valley, Sidhu Farms was established by brothers Chet and Ajmer Sidhu and their families in 1998. With certified organic blueberries as their primary crop, they also grow strawberries, raspberries, blackberries, marion and boysenberries, as well as an assortment of vegetables. In addition to these traditional valley crops, they also grow Indian squash and okra, which is sold through Indian markets in Kent and Tukwila.

 Baghwan, Mr. Ghuman, and Sharigit, Sidhu Farms, Puyallup

Sharigit and Mr. Ghuman inside Packing Shed, Sidhu Farms, Puyallup

Bhagwan Sidhu, Shinder Sidhu, Chet Sidhu, Ajmer Sidhu, Bindarjeet Kaur, Sidhu Farms, Puyallup

Bindarjeet Kaur and Amarjit Kaur Sorting Berries, Sidhu Farms, Puyallup

SEA BREEZE FARM, VASHON ISLAND

A small island farm, Sea Breeze Farm raises pastured animals for eggs, meat, and milk. Owner George Page describes his farm as"passionate about grass. We want our animals to live, love, breath, frolic, eat."

Catching Lamb for Slaughter (detail), Sea Breeze Farm, Vashon Island

 Cow Milking, Sea Breeze Farm, Vashon Island

Pouring Cow's Milk for Cheese Making, Sea Breeze Farm, Vashon Island

423
12164
423
12164

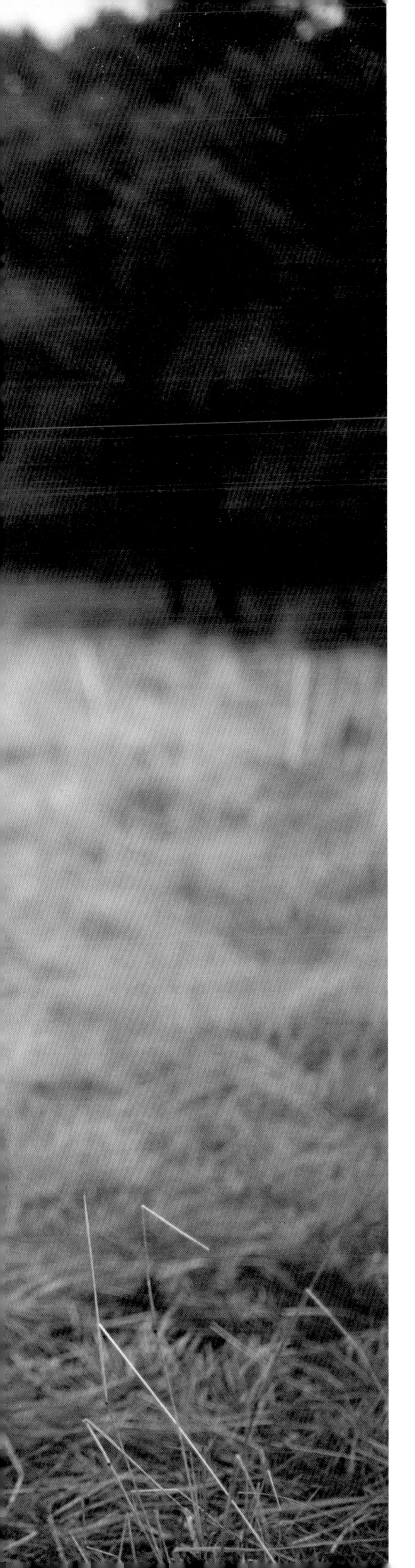

Pasture-Raised Cow, Sea Breeze Farm, Vashon Island

KAO LEE'S GARDENS, FALL CITY

Kao Lee Cha spent much of her childhood as a nomadic farmer working in the hills of Laos, China, and Cambodia. She was the first Hmong woman to get organic certification in Washington state. She farms on two acres of land with her cousin's family and has the helping hands of all of her grandchildren. Half of the farm is exclusive to growing vegetables—from golden beets and Russian kale to summer squash and Chinese spinach—while the other half is dedicated to growing flowers.

Kao Lee Cha and Grandson Kong Lor in Flower Field, Kao Lee's Gardens, Fall City

 Above and right: *Cha Family Harvesting Peas*, Kao Lee's Gardens, Fall City

CHA
28

PILCHUCK GARDENS, ARLINGTON

Pilchuck Gardens, owned by Mike Monas, is a small greenhouse tomato farm where one pickup truck provides local early variety tomatoes to the community through farmer's markets. The work is done by hand, making the greenhouses a peaceful place to be and to work.

Mike Harvesting Tomatoes, Pilchuck Gardens, Arlington

 Mike with Dog Outside Greenhouses, Pilchuck Gardens, Arlington

GROWING THINGS FARM, CARNATION

Established in 1991 by Michelle Blakely, this Certified Naturally Grown farm intentionally uses minimal machinery. Their laying flock, pastured meat birds, pastured pork, and grass-fed beef round out their produce, and contribute to the fertility of their soils.

Julie with Wheel Hoe, Growing Things Farm, Carnation

Blake with Chicken Feed, Growing Things Farm, Carnation

Blake Stocking Fresh Straw in Henhouse, Growing Things Farm, Carnation

Blake Feeding Pigs, Growing Things Farm, Carnation

JUBILEE BIODYNAMIC FARM, CARNATION

Located in the heart of the Snoqualmie Valley, Erick and Wendy Haakenson's farm has a mission, in addition to growing delicious, healthy food, to educate the public about organic biodynamic farming techniques, and to instill in future generations the value of land stewardship.

Kenneth on Tractor, Jubilee Biodynamic Farm, Carnation

BOSS

SUYEMATSU FAMILY FARMS, BAINBRIDGE ISLAND

Started by the Suyematsu family in 1928, this is the last of the large family farms that once made Bainbridge Island an agricultural community and strawberry capital. The farm highlights its sustainable and organic produce, farm-school programs, community education and local tradition-bearing. Akio Suyematsu was one of the first Japanese Americans to be forcibly removed and incarcerated during World War II; out of his experience of exclusion, his farm has become the largest and longest-operating farm in the region, and one of the most inclusive places on Bainbridge Island.

Akio Suyematsu in Raspberry Field, Suyematsu Family Farms, Bainbridge Island

 On Watch to Scare Away Crows, Suyematsu Family Farms, Bainbridge Island

OF MIND
Fremont

OXBOW FARM, DUVALL

Tom Alberg and his family established this twenty-five-acre mixed vegetable, tree fruit, and berry organic farm and education center bordering an "oxbow" lake in 1999. Natural habitat is an essential component of a healthy farm, and the farm has been working to conserve and restore habitat and to educate the community about farming and the environment.

Sarah in the Field, Oxbow Farm, Duvall

 Lettuce Harvest, Oxbow Farm, Duvall

Adam with Harvest Bucket, Oxbow Farm, Duvall

FROG SONG FARM, FIR ISLAND

Located on sixteen acres of prime farmland soil on Fir Island, Nate O'Neil and Shannon Dignum's family farm is certified organic. Heirloom potatoes are their specialty; they grow twenty different varieties, including Purple Peruvians, Yukon Golds, and Fingerlings. They use "dry farming" practices, cultivating the soil to retain and nourish the farm without irrigation.

Oliva in Flower Field, Frog Song Farm, Fir Island

 Hand-Dug Potato Harvest, Frog Song Farm, Fir Island

Oliva in Treehouse, Frog Song Farm,Fir Island

 Nate in Wheat Field, Frog Song Farm, Fir Island

From Our Own Fields: Reconnecting Food, Farms, and Communities

DR. MARCIA OSTROM

Agriculture is a central part of the Washington state economy: It drives nearly $10 billion in farm sales, boasts 37,249 farms, and uses nearly 15 million acres of land. A diversity of micro climate zones, rich soils, and ample irrigation provide ideal growing conditions for over 300 different crops, making Washington the second most agriculturally diverse state in the nation. Washington is best known for fruit orchards and wheat, producing 70 percent of all the apples grown in the United States and exporting wheat to Asia. Other top agricultural commodities include milk, potatoes, vegetables, hay, and beef.[1]

This landscape of plenty, however, belies a complex set of underlying challenges. Threats posed by climate instability, economic globalization, intensifying competition for resources, declining numbers of productive farms, the precarious status of many farm workers, and high rates of food insecurity raise questions about the long-term health of Washington's food and agricultural systems. While no simple solutions exist, burgeoning public interest in food and farming may afford new opportunities for dialogue and change. Indeed, as awareness grows about the significance of their food choices, citizens in Washington—as in many other parts of the world—are finding new ways to participate in decision making about how their food is produced, distributed, and consumed.

Over the past several decades, a rich variety of independent, local food initiatives led by entrepreneurial farmers, nonprofit groups, and passionate volunteers have taken root in the Northwest. Consumers are supporting alternative food production and distribution systems in record numbers by purchasing locally grown foods directly from farmers or by seeking them out at restaurants, food cooperatives, cafeterias, retail stores, and online. A statewide farmer survey found that roadside stands or farm stores—followed by farmers markets and you-pick farms—were the most common ways that vegetable and fruit farmers were selling products directly to consumers and the demand appears to be growing.[2] When Washington consumers were polled, over 83 percent said they wanted to buy more local fruits and vegetables. Among their top-ranked food purchasing priorities was the goal of keeping local farms in business.[3]

Nearly obsolete in Washington until a few decades ago, community-based farmers markets now draw large numbers of farmers and shoppers and are becoming increasingly popular with local governments and city planners seeking to reinvigorate local economies and community culture. Today, the state's 160 farmers markets range in size from five to one hundred farmers each, attract from 12 to 10,000 shoppers on a typical day, and have annual sales ranging from $1,000 to $5 million. Combined, these markets provide outlets for 1,200 farmers statewide. Seattle's Pike Place Market, one of the oldest continuously operating public farmers markets in the country and a major tourist destination, draws an estimated 2.5 million visits annually.[4]

Participation in Community Supported Agriculture (CSA) has also been steadily growing over the past few decades. Perhaps the most direct challenge to conventional market relationships, the CSA model asks consumers to share in the agricultural risk by paying a farmer in advance to receive a "share" of whatever is harvested each week during the growing season. In 2001, eighty known CSA farms were operating in Washington state; over a decade later, 388 farms reported sales through this channel on the U.S. Census of Agriculture.[5] CSA farms in Washington may distribute from 5 to 600 shares of food per week during the regular growing season, and offer additional optional winter and spring shares. Besides growing, harvesting, and distributing seasonal organic produce, most CSA farms offer their members education about farming and the environment, tips for cooking and preserving food, volunteer opportunities, and fun events on the farm.

A variety of other creative market relationships are evolving between local farmers and food banks, food hubs, food cooperatives, independent grocers, health food stores, restaurants, schools, and hospital cafeterias. In addition to a strong network of community-based food cooperatives, Washington is home to the largest consumer-owned natural foods retail cooperative in the country: the Puget Consumers Cooperative (PCC), an organization that maintains rigorous food purchasing standards, engages with food policy at regional and national levels, and founded a land trust to protect farmland for future generations of organic farmers.[6]

Similarly, restaurants across the state have become important allies of the region's small farms, going to great lengths to build flexible menus and storage capacity to accommodate locally available foods and extend their seasons. Seattle has led the way as a national center of culinary innovation for seasonal and local foods and farm-to-chef training programs. Around fifty area chefs belong to the Seattle Chefs Collaborative, which has a stated mission of supporting local farmers and sustainable farming practices. This group recently developed a "Road Map to a Greener Restaurant" to assist with local food procurement strategies.[7] One small farm, Quillisascut Farm, has built a dormitory, teaching kitchen, and wood-fired oven to host courses for culinary students. They learn how to butcher livestock, harvest seasonal organic crops, and incorporate them into menus.

Concerted efforts by nonprofit groups, parents, staff, students, and other stakeholders have demonstrated that it is possible for schools, hospitals, and colleges to purchase cafeteria foods from regional farmers. The Washington State Department of Agriculture reports that at least sixty farmers are registered to sell products directly to schools. Of school districts surveyed in the state, over 90 percent say they are "very or somewhat interested" in purchasing

Washington-grown foods directly from farmers.[8] Directories and "farm finder" websites, some with regularly updated fresh sheets, have been developed to facilitate local food sourcing by chefs and institutional food buyers. Recently, a number of organized efforts have emerged to more seamlessly aggregate and distribute locally grown foods to such interested buyers using a food hub model.

The pioneering efforts of the state's organic and sustainable farmers over many decades to develop productive, ecologically based farming models and the success of the alternative marketing initiatives described above are attracting new farmers. Aspiring farmers include a diverse variety of ex-urbanites, immigrants and refugees, second-career seekers, and veterans, many without family farming backgrounds or prior farming experience in this country. As a consequence, the demand for ecologically based farmer-training programs oriented toward the production of high-demand products for local markets has exploded.

While they may be proliferating across the landscape, can such disconnected, grassroots efforts succeed in realizing any degree of control or transformation of how food is produced and consumed? The rapid growth of participation by farmers, distributors, retailers, and consumers in these alternative food markets, while remarkable, cannot be assumed to further the ideals of environmental or economic sustainability, food justice, or democratic participation in the food system.[9] New marketing practices or the "localness" of a farm do not in and of themselves ensure any particular type of land or labor management practice, economic viability, or equitable food access.[10]

What are the most obvious barriers to making sustainable agriculture and access to healthy, environmentally sound and equitably produced regional foods the mainstream?

Perhaps most critical is a continuing loss of productive farms and farmland; and a loss of appropriately scaled processing, transportation, distribution, and marketing infrastructure. In Washington—as in many other highly commodity-based agricultural regions—most remaining agricultural infrastructure is oriented toward highly specialized, industrial-scale, export agriculture.

There has been a steady loss of independent, small and mid-sized processing and retail food businesses in past decades. At the same time, the health and safety regulations surrounding food processing have grown in complexity. Far more paperwork and permitting is involved in creating a value-added product to sell to your neighbor, than in selling unprocessed commodities or livestock on wholesale markets. For example, even in the most productive fruit-growing regions of Washington, it is difficult to find products like ciders, fruit pies, jams, and jellies made from local products for local retail.

On the production side, a lack of access to farmable land and water within a reasonable proximity to population centers can be prohibitive for new-entry, direct-market, and value-added farmers. Climate variability has added further production and economic risks. Finally, goals such as securing living wages for farmers and farm workers and obtaining marketplace rewards for environmental stewardship can seem to be in direct conflict with the goals of obtaining more affordable, quality foods for those who currently lack access.

The next steps for relocalizing Washington food systems will call for a new level of engagement and organizing on the part of concerned citizens.

Enacting policies to protect farmland and water for agriculture, mitigate the impacts of climate change, rebuild processing and distribution infrastructure, address regulatory barriers, support farm workers and farmers, and ensure the availability of quality food for everyone will require collective action that goes beyond making informed individual food choices. In a time of shrinking public investment in agricultural research and extension, growing economic disparities, insufficient agricultural infrastructure and looming environmental crises, identifying the most effective levers for change will be critical. Targeted citizen initiatives have already resulted in a variety of observable, long-term institutional and policy changes in Washington, however, it remains to be seen whether disparate players in regional food systems can coalesce around a shared vision and amass the organizational and financial resources needed to prompt lasting institutional and policy changes.

A new dialogue about the politics of food appears to be emerging at local, regional, and national levels. Research shows that for many Washington residents, the way they think about food has expanded beyond their personal health to include the challenges of farmers and the environment. Moreover, when surveyed, they express a belief that they can affect change through acting on their values about food at an individual level.[11] At a collective level, however, the alliances formed among producers, food businesses, environmentalists, consumers, and community food advocates can be more fragile.

Movements for change in the food and agriculture system have historically either pursued goals of improved conditions for farmers, farm labor, or the environment from the production side, or, from the consumption side, have typically addressed hunger, nutrition, or food safety issues. However, more recently, international grassroots movements of farmers and eaters have come together to promote more democratic and equitable community access to both the means of producing food and the food itself—under a banner of "food sovereignty." Regardless of the terminology or the initial entry point, food awareness and activism are on the rise both locally and globally. The challenge now is to fashion these efforts into more unified and strategic social movements.

While it is certainly true that control of the food system still largely resides with highly concentrated, powerful global actors, a new culture of food that connects people to their local farmers, ecosystems and communities is encouraging a new sense of possibility. An important measure of success for a social movement is its ability to inspire new cultural ideals and values that can be taken up and acted upon by the larger society.[12] In Washington, personal connections to food that embody values of fairness, community well being, and environmental sustainability are becoming increasingly evident. Concerted action will be required, however, to translate these values into the systems-level change needed to address current challenges and move closer to these ideals at a regional level.

ENDNOTES

1. Washington State Department of Agriculture, "Agriculture in Washington State," from United States Department of Agriculture, U.S. Census of Agriculture 2012, accessed January 15, 2015, http://agr.wa.gov/AgInWa/
2. Marcia Ostrom and Raymond Jussaume, "Assessing the Significance of Direct Farmer-Consumer Linkages as a Change Strategy: Civic or Opportunistic?" in *Remaking the North American Food System: Strategies for Sustainability*, ed. C.C. Hinrichs and T.A. Lyson (Lincoln and London: University of Nebraska Press, 2007), 235–259.
3. Marcia Ostrom and Raymond Jussaume, "Reframing Food: Understanding Trends in Consumer Food Purchasing and Implications for Agrifood Movement Mobilization in the Northwestern U.S. *Proceedings of the 10th International Farming Systems Association Symposium* (IFSA), Aarhus University, Aarhus, Denmark, July 2, 2012.
4. Marcia Ostrom and Colleen Donovan, "Summary Report: Farmers Markets and the Experience of Market Managers in Washington State," USDA National Institute of Food and Agriculture (NIFA), Agricultural Food Research Initiative, Grant #2009-55618-05172, "Engines of the New Farm Economy: Assessing and Enhancing the Benefits of Farmers Markets," 2013, http://csanr.cahnrs.wsu.edu/wp-content/uploads/2013/11/WSU-FMMS-report-Nov-2013.pdf
5. United States Department of Agriculture, U.S. Census of Agriculture, 2012, accessed May 2, 2014, http://www.agcensus.usda.gov/Publications/2012/Full_Report/Volume_1,_Chapter_2_County_Level/Washington/.
6. PCC Natural Markets, accessed January 15, 2015 http://www.pccnaturalmarkets.com/about/
7. Seattle Chefs Collaborative, accessed May 15, 2012, http://seattlechefs.org.
8. Tricia Kovacs, Washington State Department of Agriculture Washington State Farm to School Coordinator, personal correspondence, June 6, 2012.
9. Ostrom and Jussaume, "Assessing the Significance of Direct Farmer-Consumer Linkages as a Change Strategy: Civic or Opportunistic?"
10. Hinrichs, "Embeddedness and local food systems: Notes on two types of direct agricultural market."
11. Ostrom and Jussaume, "Reframing Food: Understanding Trends in Consumer Food Purchasing and Implications for Agrifood Movement Mobilization in the Northwestern U.S."
12. Melucci, Alberto. *Nomads of the Present: Social Movements and Individual Needs in Contemporary Society.* Philadelphia: Temple University Press, 1989.

This essay is drawn from a paper Dr. Ostrom presented at the "International Workshop on System Innovation towards Sustainable Agriculture," INRA, Paris, France, May 21–22, 2014

Pages 68–69: *Looking Out from Barn*, Frog Song Farm, Fir Island

Coming Home

ANNA MIA DAVIDSON

It's mystifying how you can move far away from home, only to return again—not physically but spiritually. The same force DNA has on animal migration patterns affected me and brought me to an understanding of the power of my own DNA.

I applied to college in photojournalism, but realized that following in the footsteps of my photographer father would be a long, uphill journey I'd rather not take. Instead, I created my own major: "Facilitating and Mediating Social Change." I was young and optimistic, and I wanted to change the world. While in college, I headed to postwar El Salvador with friends, despite the U.S. travel ban at the time. As my DNA necessitated, I brought my Nikon FM2 and a bag of Tri-X film. I was drawn to shoot the projects I was involved in, though I wasn't there to be a photographer; I was there to work in FMLN (Farabundo Martí para la Liberación Nacional, one of the main participants in the civil war) territory villages, helping women farm and helping the underground radio survive. But I couldn't help but bear visual witness.

When I returned to school, a friend who had a key to the journalism darkroom would sneak me in. Late at night, we would create our bodies of work. I was asked to show the images on campus as a way to get students involved in what was going on in El Salvador. And then I set my camera down for a while. It wasn't until the dean selected me as one of ten students to create an independently structured major that I decided to accompany my fifty-page senior thesis—"Women Warriors," about women student activists on the University of Colorado campus—with images. I graduated, put the handmade thesis book in a closet, and moved to the West Coast.

After training to be a union organizer, I turned down an offer for placement and instead went to Orcas Island in the San Juan Islands of Washington state, where I'd heard a permaculture farm was located. I had experience with permaculture from volunteer work I did on the Diné (Navajo) reservation, and I had always wanted to explore it further. When I reached the farm on Orcas, I knew I wanted to stay. The farmer and I agreed upon a work trade, and I set up a tent I borrowed from a guy I had met the day before. Living off the land—working with my hands, feeling wild and free—all felt right to me. When the growing season was over, it was time to decide where I would live and work next, and to begin to grow up a bit. I jumped on the opportunity to live near my sister and landed in Seattle.

There, I began working with girls who were living in a halfway house between the juvenile prison they had just left and their undetermined futures. I built a garden with them in the backyard, nurturing plants from the earth and eating our successes. The growing season ended, and I began working with homeless and disenfranchised youth. I was inspired to try and make a difference in their lives. I became the program manager within a short time, and was poised to climb up to the executive director role before I was 23 years old. But my DNA kicked in.

I started to feel trapped, like a caged bird—not wild and free.

Around this time I met a man. A burly, outdoor, mountain man who smelled of earth and hard work, who had strong hands, kind eyes, huge passion. This organic farmer completely captivated me. We had an instantaneous connection. I took him to the permaculture farm on Orcas where I had worked the summer before. We held hands the whole drive there, missed the ferry, and waited in the back of his old pickup truck in sleeping bags zipped together until the next boat came. He read me Richard Brautigan's poetry and inspiring texts by Wendell Berry. We talked about land and the freedom one feels working with the earth.

For a while, it was enough to spend my days off helping on his farm. But it wasn't long before I wanted to feel greater freedom than even the farm could offer. I remembered an opportunity I had in college to go to Cuba with an activist teacher; I regretted not going. I was once again feeling the undeniable force of my DNA. In a short time, I rented out my room in Seattle, rolled all the hundred-dollar bills I had saved into a secret compartment in the belt of my jeans, packed an old Leica M4-P, two lenses, and huge bags of Tri-X and 3200-speed film in a small backpack, and headed to Cuba to photograph for a few months.

When I returned to the United States, I went home to New York City, and to my father's darkroom. Like a penguin returning to feed in the sea, I was drawn to the roots of my birthplace. I spent days printing archival fiber prints from the trip to Cuba, and when I saw them, I realized that I had made something. I took the prints back to Seattle, and applied for the Eddie Adams Workshop, a rite of passage for many seriously interested in photojournalism. After I got accepted, I was informed that the workshop would be shot on slide film, a medium I had never used. I rushed to the store, bought several rolls, headed to my boyfriend's farm, and learned all about slide film, and the strength of love.

As a result of my experiences and connections through the Eddie Adams Workshop, I landed a freelance job shooting in Seattle with the Associated Press. I learned quickly how to secure contacts and relationships directly with editors from papers around the world, which enabled me to start shooting for newspapers and magazines directly. The Blue Earth Alliance granted me sponsorship to develop my photographic project in Cuba; Kodak gave me a generous film grant; and I saved up enough money to return to Cuba several more times over the next ten years to complete the project.

During that same decade, the farmer, John, and I got married, and later had a baby. With the birth of our daughter, Eden, I truly learned the depth of love one can feel for another, and I made what felt like an easy decision to primarily be at home to raise her. For an independent lover of freedom, this was a big decision. I watched as John continued on with his life as normal, wild and free on the farm even in the winter, while I was in our home in the city with our infant child. I loved attaching myself to Eden—she was rarely seen off of my body. We would walk and talk and roam as one, and by 18 months old she was speaking in sentences. "An old soul," people would say, as her wise eyes humbled many who looked into them. From a young age, Eden loved being on the farm, and I could feel a sense of freedom when out on the land with her.

It was not until Eden turned two that I realized I needed to photograph again. I made one last trip to Cuba, this time with John. I wanted to show him how incredible Cuban agriculture practices were, and I wanted to be certain that my

project was complete. A part of me also hoped for a reconnection of the energy we felt when we met, two free spirits who shared a love for the outdoors, adventure, the earth, and agriculture. I realized that my project was officially complete when I felt like I was starting to retake the same images. The Cuba I had experienced ten years earlier had changed, and I wanted to shoot closer to home. On that return flight, I realized I had once again traveled far away, only to return home again—this time home to my roots in agriculture, home to my passion for the earth and growing food. I had been seeking what was in front of me all along.

Over the years I had dated John, at times I had made photographs on the farm, capturing the beauty of life as it unfolded. But it wasn't until the last trip back from Cuba with him that I decided to be deliberate about it—to shoot in color, and in medium format, using an old Hasselblad and a Rolleiflex my father had given me.

After nearly two decades of loving a farmer and embracing the life we shared together as a farm family, I was used to long days working on the farm and many evenings discussing our vision as well as the farm as a business. I loved the idea that I could marry two of my passions together—photography and farming—and use photography to document the life and people I loved. The sustainable farming life for me is so much more than just a way of life. It feels like a religion, a spiritual calling to tend to the earth as a steward of the land, giving back to the earth as well as reaping to keep things in balance. As attracted as I was to that handsome rugged farming man I married, I was equally drawn to the movement of sustainable farming and what it means for local communities—and for the globe.

As I was reaching the end of my rolls of film, I realized that the reality of the farm existence was that cash always seemed to be scarce, and I may have to put this project on hold. Shortly after I put my funding needs into the universe, I was approached by Michelle Dunn Marsh, who through the Aperture Foundation nominated me for a commission from USA Network to photograph the "Character of America." I had once shown Michelle a few photographs from the origins of my farm project, and it was ultimately a photograph of John (page 5), that captured her attention. Having grown up in a farming community in the Pacific Northwest herself, she felt she was a kindred spirit to the vision of my work, and wanted to see the project move forward.

I took it as a sign that this project must go on, and when I was selected for the USA Network project I was fortunate to also receive a Kodak film grant to shoot 120, so I could continue to use my medium-format cameras. For several weeks I roamed the land, approached farmers at farmers markets, and set up appointments to meet them on their land to photograph them. Many of the farmers I knew well; some I did not know at all, but they all knew our farm, Nature's Last Stand. I was seen as a farmer, a farmer's wife, and a member of the farming community. I sometimes had Eden on my back, taking walks along the farm road while shooting on our land. Farm interns, employees, and

 Eden Collecting Raspberries, Nature's Last Stand, Fall City

John himself became the focus of my photographs during long harvest days.

We decided to rent out our house in Seattle and live on the farm for the summer, a plan that would inspire us to do that for three months every year so that we could have more time together as a family while John worked long hours on the farm, and so our children could grow up connected to the earth. I loved simplifying, living with no electricity or running water, bathing in the river, swimming naked, planting seeds, pulling weeds, growing vegetables, watching piglets be born, collecting eggs, cooking the food we grew or bartered. Eden declared that her favorite breakfast was fish caught in the early morning with Daddy, accompanied by freshly picked wild fiddlehead ferns and salmonberries. At two years old, she would help us harvest by trailing behind the bucket, nibbling a mousy bite out of each vegetable; at three years old, she would run across the field alone to find Daddy, with a great sense of power, independence, and freedom, her little legs shorter than the growing vegetables, like splashing waves of green, around her.

The Aperture/USA Network project involved not only making new photographs but seeing them published in a book and exhibited in a five-city traveling exhibition. I felt grateful for the opportunity to dive so deeply into a personal project. At about the time the project concluded, we were blessed with the birth of our son, Elias.

With a summer baby, I once again found myself focused on being a mother, knowing now that infant time was fleeting and I wanted to savor it. Though I have always been a strong, independent "woman warrior," getting things done and paving a course in the predominantly male field of photojournalism, I found I also took pride in being nurturing, loving, and somewhat traditional. I didn't know that about myself until I had children.

The first day home from the birthing center, with our new infant son on my breast and Eden in my lap, I was sitting on the floor of our house when John left me there alone to go to the farm and feed the pigs. I felt abandoned in that moment—overwhelmed by two helpless lives entirely dependent on me. I thought, "how in the world am I going to do this?" My role models were male photographers of my father's generation—they did not raise the children, they roamed the earth creating their photographic work and left the work of raising children and running households to others. There were women photographers of that generation who didn't have children; I could look to them only as great examples of women making it in a male-dominated field, not as guides for how to be mother-photographers. Then I thought of Sally Mann, a well-loved name in my parents' household. They greatly admired her work, I think because they knew how complex it was to have adolescent daughters. That thought was a comfort.

In the early months of Elias's life, we lived on the farm. As I had done with Eden, Elias was wrapped on me as I worked in the field. Eden would trail along closely. We picked

Elias and John, Nature's Last Stand, Fall City. The freshly foraged cumfrey plants are to feed the pasture-raised pigs.

blackberries, took walks, and harvested food for our dinner. We helped where we could on the farm in between nursing, snuggling, and tending to domestic chores.

The next summer, I was contacted by Fotodocument, an organization interested in commissioning photographers internationally who were documenting positive environmental initiatives. I was starting to see that sustainable food production was not just a topic of interest locally, but a global issue. I began to see my life on the farm and being married to a farmer as part of something greater, and I picked up my camera again. Shortly thereafter, the city of Seattle chose eight of my images for billboards across from City Hall to be showcased for two years; the people working in the offices there requested they stay up an additional two years because they liked seeing the environment and farmers outside their windows. Somehow seeing the earth and stewards of the land was hopeful and promising within the concrete jungle.

• • •

I grew up going on photo dates with my dad, Bruce Davidson. We'd head out on the streets, shoot a roll of black-and-white film, and go back to his darkroom in our apartment to develop the negatives and make prints. Pure magic. We'd make pinhole cameras, and I'd photograph my dollhouse dolls, making their portraits for the dollhouse walls. He taught me how to approach people to ask their permission to take photographs, and then how to turn around and make myself invisible. He taught me patience in waiting for the right moment—when the person forgets you are there or lets down their guard and connects with you in a deeper way. With a sixth sense, a sensitivity, an awareness of others—a photographer like my father was not so different from what, as an activist, I try to do in shedding light and bringing awareness.

My father's large prints from various bodies of work such as *East 100th Street*, *Subway*, *Brooklyn Gang*, *Civil Rights*, or individual iconic photographs like *Welsh Pony*, and the Verrazano-Narrows Bridge lost in the graininess of fog, surrounded me. Starting at a young age, I was also exposed to influential photographs by Sebastião Salgado, Burt Glinn, Josef Koudelka, Diane Arbus, Elliott Erwitt, W. Eugene Smith, Susan Meiselas, Larry Towell, Robert Capa, Sally Mann, Marc Riboud, Robert Frank, Dorothea Lange, and Henri Cartier-Bresson. Their images were windows into different worlds that I had not yet seen, or might never see. Our New York City apartment was often a crash pad and welcoming stop for friends from faraway lands—photographers, artists, poets, and playwrights. Their books graced our shelves, and we attended their exhibition openings. It was an endless tour of photography, a river of images running through my developing mind.

I remember late nights at the photo festival in Arles, France, when photographs were projected outside on an ancient wall while my parents drank wine and socialized late into the night—noise and visuals washed over me, bathing me in influence at eight years old. There was a contagious energy, passion, and freedom all around me. There were workshops around the world, where my father would teach and lecture, often bringing my mother, sister, and me. There was the special time when I was nine that he invited me alone to the Maine Photography Workshops.

My father home schooled me in photography, and through that, he taught me how to navigate life. Survival was about having duct tape, a pocket knife, a handkerchief, street smarts, and the ability to relate to people in tough situations. My mother inspired a sense of importance on humbleness, intelligence, grace, and hospitality, and on being a strong, independent, feminine, sexy woman. Throughout my childhood, she was an active artist in her own right: an actress, model, and writer. She brought a nurturing energy as well as a love of arts and culture into our home.

The topic and business of photography was a cloche that surrounded our family at all times—sometimes inspiring and comforting, while at other times suffocating. The studio was run out of our home, incoming business calls were taken at the dinner table, and nothing felt as sacred as the work created. The overall message throughout my upbringing was that success was driven by what you create, not by the money you make. The photographic world that surrounded us has had a profound influence on the lens through which I look at life. I set out to pave my own way and follow a different path, but the strength of DNA is powerful.

• • •

I believe that photography has the unique ability to capture the essence of a person's being. A photograph can bear witness to events, inspire and motivate people into action, and be a powerful tool for social change by allowing viewers to digest important truths through the visual experience of these simple, quiet moments. Photographing for this project was for me a personal journey into the lives of each of the farmers, and each experience was inspiring. The amount of work and dedication required to be a true steward of the land is humbling to witness. Farmers exemplify a way of life that is often overlooked, yet they represent a universal experience that is integral to the fabric of our world.

Agriculture is a link we share with practically every other nation on earth and a bond that makes us more similar then different. At its core, agriculture is universal, not only because it feeds us, but because it feeds our sense of wonder in the natural world. Despite their cultural, educational, and familial differences, the farmers I photographed in the Pacific Northwest are linked by a movement of sustainability that is larger than each individual.

I originally wanted our farm to be called "Uprising Farm." It invokes a wild and free feeling in me, like a sprout about to bust through the earth to reach the light. But we agreed that John would get the final word on the farm name, and he chose "Nature's Last Stand." In the end, I think his choice speaks to what is happening to the environment today. The earth is speaking to us against the confines of monocropping, factory farming, pesticides, and herbicides—standing against Monsanto seed control and environmental contaminants.

Our future depends on a new appreciation for local production and consumption. We will need to embrace what it looks like to go back to our roots, and I hope through this project to inspire our nation to take a drastically different approach to our food system. Though powerful social and political influences can lead us astray, the individuals in these photographs have restored my optimism in the human spirit and what we are capable of creating and protecting. I am thankful that we still have invaluable farmland in the United States—what is left needs to be protected. A balance of coexistence between humans and nature is at the root of the solution.

Biographies

ANNA MIA DAVIDSON

Photography has been Anna Mia Davidson's passion for more than two decades, both as a tool to document diverse cultures and environmental issues, as well as a medium for social change.

An award-winning photographer, her work on sustainable farmers in the Pacific Northwest has been supported by a grant from Fotodocument.org; selected for long-term exhibition by the city of Seattle; and was featured in the March/April 2014 issue of *American Photo* magazine. It was supported by and debuted on the USA Network, was featured on the Today Show and was published by Chronicle Books as part of *The Character Project* in 2009, a project coordinated through the Aperture Foundation.

When she is not working on her own projects, she is a freelance photographer for the Associated Press as well as a variety of news and magazine publications such as *Vanity Fair, The New York Times, Christian Science Monitor, Reader's Digest, New York* magazine and *The International Herald Tribune*, and various commercial clients.

She lives with her husband and two children in Seattle, where she also serves on the board of Photographic Center Northwest.

SEBASTIÃO SALGADO

Brazilian photographer Sebastião Salgado is one of the most respected photojournalists working today. Appointed a UNICEF Special Representative in 2001, he has dedicated himself to chronicling the lives of the world's dispossessed, a work that has filled ten books and many exhibitions and for which he has won numerous awards in Europe and in the Americas.

Educated as an economist, Salgado began his photography career in 1973. His first book, *Other Americas*, about the poor in Latin America, was published in 1986. This was followed by *Sahel: Man in Distress* (also published in 1986), the result of a fifteen-month-long collaboration with Medecins San Frontières covering the drought in northern Africa. From 1986 to 1992 he documented manual labor world-wide, resulting in a book and exhibition called *Workers* (Aperture, 1993), a monumental undertaking that confirmed his reputation as a documentarian of the first order. From 1993 to 1999, he turned his attention to the global phenomenon of mass displacement of people, resulting in the internationally acclaimed books *Migrations* and *The Children* (Aperture, 2000).

Working entirely in a black-and-white format, Salgado's respect for his subjects and his determination to draw out the larger meaning of what is happening to them has created an imagery that testifies to the fundamental dignity of all humanity while simultaneously protesting its violation by war, poverty and other injustices. His most recent book and exhibition, *Genesis* (Taschen, was featured at the International Center for Photography in New York in 2014. Mr. Salgado lives in Paris, France, with his family. His wife, Lélia Wanick Salgado, directs their company, Amazonas Images, and has designed his major books and exhibitions.

MATT DILLON

Matt Dillon is owner and chef at the Sitka & Spruce restaurant in the Capitol Hill neighborhood of Seattle, Washington. His other food ventures include the Corson Building, Bar Sajor, bar ferd'nand, the London Plane, the little London Plane, and Old Chaser Farm. He is a 2012 James Beard Award winner in the Northwest category; in 2007 he won *Food & Wine* magazine's best new chef award. A proponent of local and seasonal cooking, Dillon is a former professional forager known for cooking with nettles, fiddlehead ferns, and morel mushrooms.

DR. MARCIA OSTROM

Marcia Ostrom is an Associate Professor in the School of the Environment and the Director of the Small Farms Program in the Center for Sustaining Agriculture and Natural Resources at Washington State University. She is passionate about food as a way to reconnect people, culture, and nature; her original essay for this book draws from her recent research.

She teaches courses and leads interdisciplinary research and extension projects to build understanding and improve the sustainability of Washington's food and farming systems, and has developed educational programs for beginning and immigrant farmers on values-based farming and marketing strategies; ecologically-based farming techniques; and production for regional food systems.

Her academic work focuses on movements for change in the food and agriculture system at community, regional, national, and international levels. She has a Ph.D. in Environmental Studies from the University of Wisconsin, an M.S. from the Cornell College of Agriculture and Life Sciences, and an A.B. degree from Harvard.

Special Thanks to

OXBOW ORGANIC FARM AND EDUCATION CENTER

LEICA AKADEMIE

NEIGHBORHOOD FARMERS MARKETS

and the rest of our co-publishers:

Stefanie Arneson
Kathleen Atkins
Subhankar Banerjee
Ronny Bell
David and Sandy Berler
Bill Bowden
Constance Brinkley
Elizabeth Brooks
Adam C. Butcher
Wendy and Tom Byrne
Gigi Casey
Michele Catalano
Jim Cox Chambers

Della Chen
Serena Connelly
Kathy Creahan
Scott Cuming
Chris Curtis
Bruce and Emily Davidson
Jenny Davidson
Gregg and Sara DePonte
Jesse Diamond
Kevin and Cecilia Dunn
Claire Dyckman
Karen Genda
Rosalyn Gerstein

Karaminder Ghuman
Melissa Gordon
Howard Greenberg
Carol Grossmeyer
Wendy Haakenson
Blaine and Joan Harris
Joan Herman
David Hilliard
Michael Hoeh
MS Holland
Aaron Holm
Janet Holmes
George and Gail Huschle

John Jenkins III and
Stephen Lyons
Eirik Johnson
Cameron Karsten
Todd Kelly
Jeffrey Klotz
Tracy Krauter
Megan Kroh
Laureen Lazarovici
Cissy Lefer
Lisa Leone and
James Bollettieri
Leah Levine

Andrew Lewin

Karen Lewis

Kayla Lindquist

Douglas Ljungkvist

Claudia Loeber

Jessamyn Lovell

Craig Macleod

Alistaire Fortune Marsh

Brian Marsh

Robin Wehl Martin

Silvia, Christian and Madison McIntyre

Nion McEvoy

Beth Mondzac

Martin Morfeld

Amy Osborne

Lily Ostle

Joseph Park

Janey Petty

Gloria Pfeif

Sylvia Plachy

Sally Poliak and Nance Tourigny

Mayumi and Lincoln Potter

Matt and Amy Ragen

Ritu Rathi

Bob Redmond

Becky Reimer

Andrew Renneisen

Sue Roundy

Anastasia Samuelsen

Beth Schilling

Rick Smith and Gayle Greene Smith

Tom A. Smith

Patricia Snavely Vogel

Seana Steffen

Jennifer Stuart

John Stevenson and Cynthea Bogel

Robert Tommervik

Matthew and Deanna Tregoning

PCC Farmland Trust

Anastasia Dunn Van Dyke, Steve and Otto Van Dyke

Sara Waldron

Cass Walker

Patricia Waterston

Adam L. Weintraub

Mick Woynarowski

ACKNOWLEDGEMENTS

In life we are lucky if we get to walk with people by our side who lift us up and help propel us forward. I would like to thank a few of the people who have been instrumental in supporting me, and the creation of this book.

My endlessly supportive parents Bruce and Emily Davidson have encouraged and inspired my journey as a human being in this world. They have stood behind the choices I've made and have been the greatest advocates of my photographs.

I would also like to thank John Huschle, a true steward of the land, who for the past nineteen years has inspired me with his passion for the earth. His profound work ethic gives a new meaning to "blood, sweat, and tears." As a lover, best friend, and husband, John introduced me to the beautiful world printed within the pages of this book, and for that I am eternally grateful. My life and soul is enriched because of the years spent on our farm. I am also eternally thankful for our children, Eden and Elias, who have shown me a depth of love I never knew existed. Watching their pure joy and love of the earth and farm life inspired me to create many of these images.To witness their carefree freedom and confidence, which has come from getting dirty, helping with jobs that needed to be done, climbing trees, finding the earth's natural treasures, and delighting in eating food growing out of the ground has been life changing for me.

I have gratitude for the land that has sustained my family, providing food for us and for our community. A huge thank you to the fellow farmers who graciously opened up their world and their lives to me, for their trust and precious time during the busiest part of their season.

No words can express the gratitude I have for Sebastião Salgado, a personal inspiration in my life. To see his passionate words printed in this book will forever bring me joy. Thank you also to Matt Dillon, to EuJean Song for assisting with Matt's participation, and to Dr. Maria Ostrom. As a photographer, it is imperative to have allies in our corner who believe in the work we do—it's how we move our images from ideas into reality. Supporters include Kodak, USA Network, Aperture Foundation, Fotodocument, City of Seattle Office of Arts and Culture, Artist Trust, and Seattle Public Library. A special thank you to all the co-publishers, especially Gretchen Garth, Katherine Anderson, Tom Smith, Melissa Campbell and Chris Curtis, and to Michelle Dunn Marsh and Steve McIntyre of the Minor Matters team for believing. To all who have been on this journey with me, I share this book with you.

—Anna Mia Davidson

Library of Congress Control Number: 2015901638
ISBN: 978-0-9906036-3-4

Human Nature is published by Minor Matters:
Michelle Dunn Marsh, *founder and picture-slinger*
Steve McIntyre, *partner and platform chief*
Co-publishers listed on pages 78–79

Developmental editing by Giselle Smith
Color files prepared by Juan Aguilera, Photographic Center Northwest, and by Art and Soul, Seattle
Printed by Oddi, Iceland

Minor Matters is a publishing platform for contemporary art, bringing books into being in collaboration with our audience. We focus on work that articulates the surface of life, bringing insight and cadence to the worlds we occupy.

To learn more please visit **www.minormattersbooks.com**
Minor Matters Books LLC—Seattle, WA. (206) 856-6595
info@minormattersbooks.com

FIRST EDITION 10 9 8 7 6 5 4 3 2 1